手牵手

Hand in Hand

Gunter Pauli

[比]冈特·鲍利 著

[哥伦]凯瑟琳娜·巴赫 绘

唐继荣 译

上海远东出版社

丛书编委会

特别感谢以下热心人士对童书工作的支持：

匡志强　宋小华　解　东　厉　云　李　婧　庞英元
李　阳　刘　丹　冯家宝　熊彩虹　罗淑怡　旷　婉
杨　荣　刘学振　何圣霖　廖清州　谭燕宁　王　征
李　杰　韦小宏　欧　亮　陈强林　陈　果　寿颖慧
罗　佳　傅　俊　白永喆　戴　虹

目录

Contents

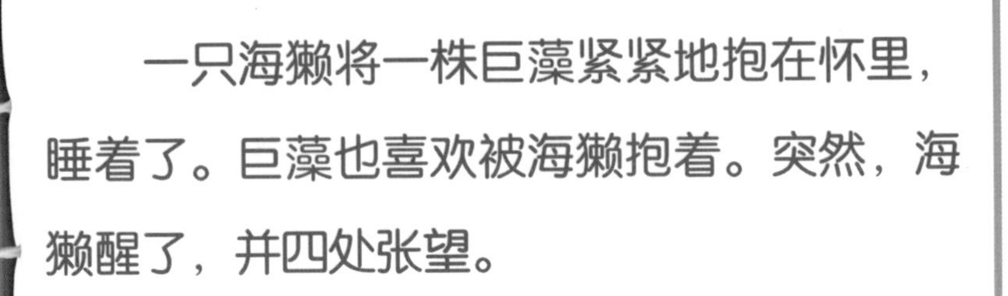

一只海獭将一株巨藻紧紧地抱在怀里，睡着了。巨藻也喜欢被海獭抱着。突然，海獭醒了，并四处张望。

A sea otter is holding a branch of kelp tightly in her arms. She is asleep. The kelp just loves to be embraced by the otter. Suddenly the sea otter wakes up and looks around.

喜欢被海獭抱着

loves to be embraced by the otter

早上好

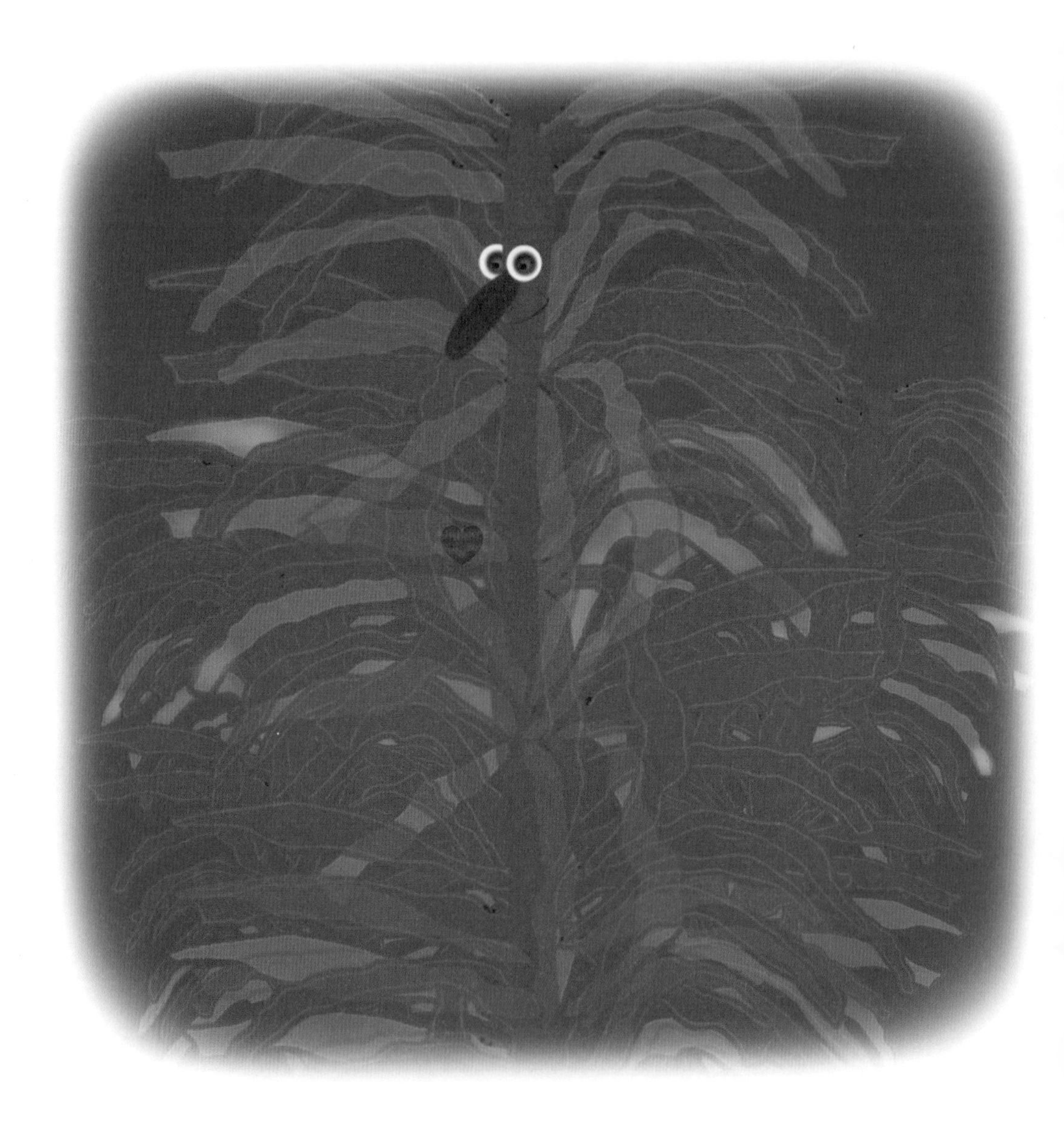

Good morning

“早上好，”巨藻说，“亲爱的朋友，我期待与你度过每一天，这让我保持健康和快乐。”

“噢，醒来后能听到一个可爱的声音告诉我说，我在把自己的事情做好时，也能帮助到你，这真是太好了！”海獭高兴地回答道。

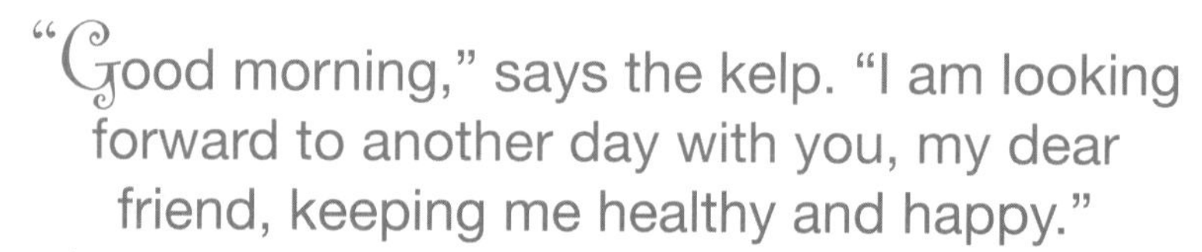

“Good morning,” says the kelp. “I am looking forward to another day with you, my dear friend, keeping me healthy and happy.”

“Oh, it is so nice to wake up and to hear a lovely voice telling me that I am doing good for you by doing good for me,” responds the otter.

“你做的事情对我的帮助可大了！如果你不把海胆、海螺、鲍鱼、海蟹甚至鱼类吃掉，我将永远无法生长成林！”

“我得承认我的胃口确实有点大。为了跟上我快节奏的生活，我每天需要吃自己体重四分之一的食物。”

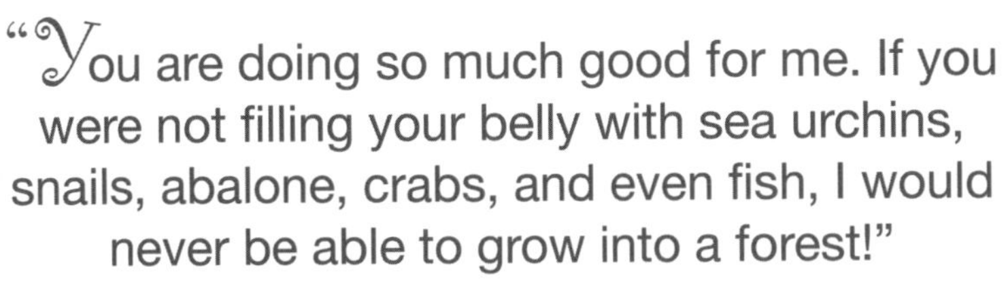

“You are doing so much good for me. If you were not filling your belly with sea urchins, snails, abalone, crabs, and even fish, I would never be able to grow into a forest!”

“I must admit I do have a good appetite. In order to keep up with the fast pace of my life, I need to eat a quarter of my body weight every day.”

我的胃口确实有点大

I do have a good appetite

我提供一个家……

I provide a home...

“是啊，你的胃口帮助我分散巨浪的能量，保护海岸带免遭风暴冲击。我能为海葵、海绵和鱼类提供一个家，甚至为人类提供食物，为你提供睡觉的安乐窝。”

“Well, your appetite helps me to spread the power of powerful waves, to protect the coast from the pounding of the storms. And I can provide a home to anemones, sponges and fish. I can even give food to people and offer you a place to sleep.”

“有人说你是关键物种，因为你利用来自太阳的能量生产食物。而且，你不仅在海洋中提供能量，在陆地上也同样提供。”

“好吧，你见过由于吃了我被冲上岸的叶片上的大量虱子而变得非常肥胖的鸡吗？这些鸡几乎难以行走，因为他们真是太重了！”巨藻笑着说道。

“Some say that you are a key species because you convert energy from the sun into food. And you do not give energy only in the sea, but also on land.”

“Well, have you ever seen such fat chickens, eating the tons of lice found on my beached fronds? These chickens can hardly walk because they weigh so much!” laughs the kelp.

你见过非常肥胖的鸡吗？

Have you ever seen such fat chickens?

……让他们的冰激凌口感柔软

...make their ice cream soft

“人类把你加到他们的冰激凌和牙膏中，这是真的吗？”

“是真的呀！人类这么需要我，是不是棒极了？虽然他们已经尝试过了，但还未发现哪种化学物质能像我一样让他们的冰激凌口感柔软。”

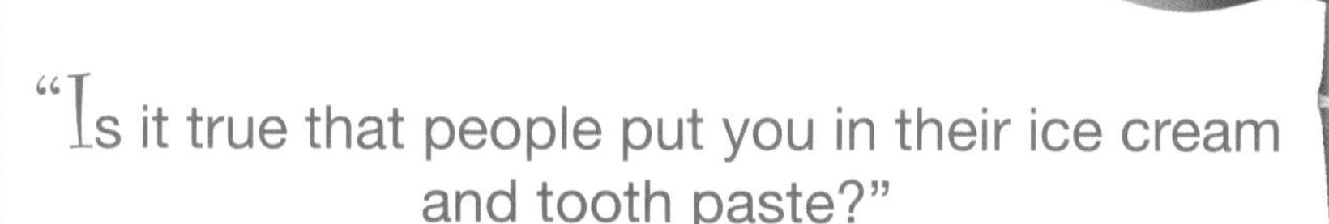

“Is it true that people put you in their ice cream and tooth paste?”

“Yes, isn’t it wonderful that people want me this much? Even though they’ve tried, they have never found a chemical that can make their ice cream taste as soft on the palate as I do.”

“嗯，人类喜欢我柔软的皮毛，并用它覆盖他们的皮肤。”

“那样柔软的皮毛几乎让你失去生命。你有最厚实的皮毛！人们为了得到它，将捕捉并杀死你们全家。”

“你知道，人类甚至开始计算我每平方厘米有多少毛发，但他们数不清楚。”海獭笑着说道。

“Well, people love my soft fur. They love to cover their skin with my skin.”

“That softness has nearly cost you your life. You have the thickest fur ever! People will hunt and kill your whole family for it.”

“You know, people once even started counting how many hairs I have per square centimeter, and they lost count,” laughs the sea otter.

人类喜欢我柔软的皮毛

People love my soft fur

……利用工具来获取你们的食物

...use tools to get your food

“每当我见到你，我就想起黑猩猩。”

“算了吧，我看起来一点儿也不像黑猩猩！我觉得自己可爱多了。”

“你和黑猩猩都是可爱的哺乳动物，都能利用工具来获取食物。这就意味着你们像人类一样聪明。”

“嗯，你还指望我不用工具就能破开贝壳吗？如果我试着去把贝壳咬碎，会把自己牙齿弄断的。”

“我最喜欢你的一点就是你具备良好的行为举止，用餐结束后，你会洗手和刷牙。”

“Whenever I see you I think of chimpanzees.”

“Come on, I look nothing like a chimpanzee! I certainly think I am much cuter.”

“Both you and the chimp are lovely creatures, both are mammals, and both use tools to get your food. That means you are as smart as humans.”

“Well, how else do you expect me to crack shellfish open? I would break my teeth if I were to try and crunch them to pieces.”

“What I love most about you is that you have good manners. You wash your hands and brush your teeth after you have had a meal.”

“你今天给了我这么多的赞美！难怪我喜欢蜷缩在你的叶片里。”

“你这样说让我很高兴！但我的确知道，你更喜欢与你的家人手牵手一起睡觉。”

“当然啦！如果你身边有这样一群家人和朋友，难道你不喜欢和我一样手牵手吗？”

……这仅仅是开始！……

“You have so many compliments for me today! No wonder I just love to roll in your fronds.”

“It is so nice of you to say that to please me, but I do know you prefer to hold hands with your family while sleeping.”

“Of course, would you not like to do the same, if you were surrounded by such a wonderful troop of family and friends?”

... AND IT HAS ONLY JUST BEGUN!...

……这仅仅是开始！……

... AND IT HAS ONLY JUST BEGUN! ...

Did You Know?

你知道吗？

Sea otters do not have fat (blubber) to keep them warm, instead they have the densest fur of all animals in the animal kingdom, with more than 100 000 hairs per square centimeter to insulate them against the cold. That is up to 800 million hairs for an adult male sea otter.

海獭并没有脂肪来为它们保暖，但它们有动物界最致密的皮毛，每平方厘米有多达 10 万根毛发，以此来保暖。这就是说，一只成年雄海獭有多达 8 亿根毛发。

Sea otters and kelp are both keystone species, meaning that their role and impact in their ecosystem is greater than other species.

海獭和巨藻都是关键物种，这意味着它们在其生态系统中的作用和影响力比其他物种要大。

Sea otters have a rich and varied diet, eating about 40 different marine species. Their great appetite spurs the growth of kelp forests.

海獭的食性复杂，能吃约 40 种不同的海洋物种。它们的大胃口能促进巨藻林的生长。

Orcas and great white sharks are predators of sea otters.

虎鲸和大白鲨是海獭的天敌。

Kelp forests are one of the most productive and dynamic ecosystems on Earth, and are recognised as a great fixer of carbon dioxide in the sea.

巨藻林是地球上最有生产力和活力的生态系统之一，并被认为是海洋中的一个大型二氧化碳吸收器。

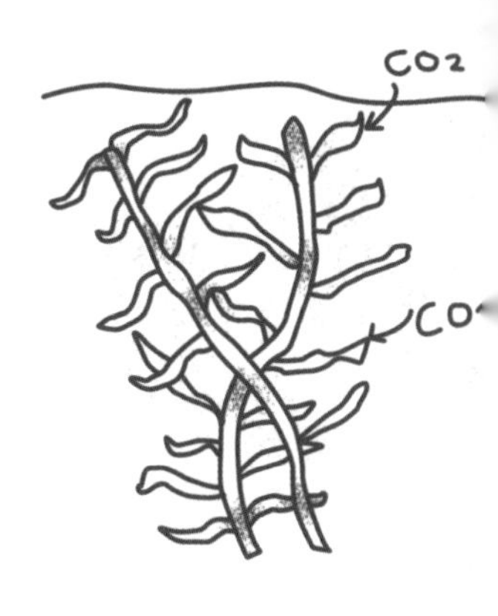

Kelp is used in ice cream, dressings, cosmetics and toothpaste, and as an ingredient in preparations that make people lose body weight. It is even used for water-and fireproofing fabric.

巨藻被用于制作冰激凌、调料、化妆品、牙膏，也是减肥产品的成分之一，甚至被用来制作防水和防火纤维。

Sea otters do hold hands while sleeping. After every meal they feverishly clean their teeth and wash their hands and fur.

海獭在睡觉时手牵手。每次用餐结束，它们狂热地刷牙、洗手、清洗皮毛。

By 1911, when commercial fur trade was banned, the worldwide population of sea otters had dropped to perhaps only 2 000 animals. The Russian population of sea otters is now the largest in the world.

当商业性皮毛贸易于1911年被禁止时，全世界的海獭数量已经下降到只有2 000只了。目前，俄罗斯的海獭数量是世界上最多的。

Think about It

想一想

Could
you ever have imagined that there are animals that hold hands and brush their teeth, naturally?

你是否想象过有不经人训练就会手牵手并刷牙的动物？

Would
you have learned how to brush your teeth and wash your hands if your parents had not insisted on it for over and over again?

如果父母没有坚持让你去做的话，你能否学会刷牙、洗手？

How
would you look if you had a hundred thousand hairs on each square centimeter of your scalp?

如果你每平方厘米的头皮上有10万根头发，你看上去会怎样？

Do
you think it is funny that the great appetite of the sea otter makes kelp grow into a huge forest, perhaps as big as the rainforest?

海獭的大胃口使巨藻长成巨大的海藻林，或许像热带雨林一样大。你是否认为这很有趣？

Do It Yourself!

自己动手！

Let us make an ice cream, the simple way. A recipe may say you need milk, cream, sugar and vanilla and even a touch of salt. We like to make an ice cream the easy way, and only use a banana. It is vegan, gluten-free, dairy-free and has no added sugar. And thanks to the richness of pectin, there is no need to add alginates that usually keep the ice cream soft on the palate. I am sure that everyone who tries it will love it.

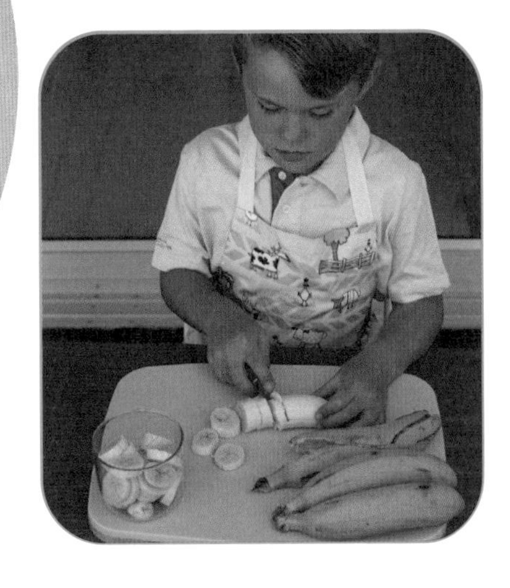

让我们用简单的方法制作一个冰激凌。食谱上可能说，你需要牛奶、奶油、糖、香草，甚至一些盐。但我们想用简单的方法来制作冰激凌，只用一根香蕉。这是一种纯素食，不含麸质或奶类成分，不添加糖分，由于香蕉富含果胶，也不需添加使冰激凌口感柔软的海藻酸盐。我相信每个品尝过的人都会喜欢它的。

学科知识

Academic Knowledge

生物学	海洋中的藻类数百万年基本保持不变；巨藻是世界上生长最快的植物，可以长达40米；海獭通过触觉发现黑暗中的猎物，从肛腺分泌出刺激性气味来自卫；被冲上岸的巨藻会吸引虱子，鸟类捕食虱子，形成食物循环；巨藻林的生物多样性像热带雨林一样高。
化 学	微生物发酵的海藻酸钠至今还无法与海藻生成的天然产品媲美；巨藻富含氮、钾，是极好的肥料，广泛用于马铃薯种植；巨藻含有碳酸钠和碳酸钾，用于玻璃制造和肥皂生产；巨藻可以提供丰富的碘。
物 理	冰激凌的柔软度由空气和海藻酸钠来决定；海獭的代谢率较高，以便应对寒冷环境下的热量损失；海獭的肾脏能通过一个类似于反渗透的过程从海水中提取淡水并产生浓缩尿，所以它们可以喝海水。
工程学	牵手对于海獭来说，类似于锚对船的作用：这是一种避免在海上漂走的方式；通过在表面创造只有0.1毫米大小的气穴来实现柔软的效果；反渗透是一种从海水（或被污染的水）中生产饮用水的方法。
经济学	对天然产品的过度要求导致资源枯竭和物种灭绝，因此迫切需要进行市场监管。
伦理学	过度利用后，由于自然资源本身不能及时更新，导致供应减少、价格上涨，进一步刺激资源破坏，我们怎么能盲目追随市场呢？
历 史	海獭和海狸在19世纪的毛皮贸易中处于中心地位；在长达4 000年时间里，巨藻是爱尔兰和苏格兰经济成分之一，并在1946年的马铃薯饥荒时期成为一种替代食物；在19世纪，用巨藻生产的沼气被用于工厂照明。
地 理	世界上最大的巨藻林沿着冷水海岸带分布；一条连绵的“巨藻公路”从日本延伸开来，向北沿着西伯利亚经过白令海峡到达阿拉斯加，向南沿着加利福尼亚海岸线直达厄瓜多尔。
数 学	海獭能在15秒内击打鲍鱼壳45次；为了成功吃到鲍鱼肉，它必须潜水4次，每次1—4分钟，其间面对冰冷的海水，它必须保持体温。
生活方式	降低对那些用穿动物皮毛显示财富和地位的人的社会认可度。
社会学	由于对动植物生存的负面影响，文化如何激励人们舍弃那些曾在过去被认为是身份地位象征的产品；手牵手一起走是一种表示亲密、友谊和信任的传统。
心理学	个人的行为与生命网络联系在一起，看上去孤立的事物实际上并非如此：自然界所展示的友爱与呵护会对我们的情绪产生影响。
系统论	生态系统的制约与平衡：海獭控制巨藻的捕食者，这样巨藻就能沿着海岸带生长成林，进而保护海岸带免遭潮汐、风暴影响。

情感智慧

Emotional Intelligence

巨　藻

巨藻享受与海獭共处的时光。他意识到海獭家族带来的帮助，并对此非常感激。如果没有海獭的辛劳，巨藻家族不可能像森林那样繁盛。巨藻解释说，巨藻林能保护海岸带免遭风暴冲击。巨藻进一步分享信息，他被冲上岸的叶片吸引虱子，而虱子是鸟类重要的食物来源。巨藻显示出自信，指出即使经过多年研究，但还没有人工合成的替代物能代替他的作用。巨藻也富有同情心，充分意识到为了获得海獭皮毛而进行的捕杀导致海獭灭绝的危险。然后，谈话变为赞美，巨藻表达了对海獭手牵手这一行为的赏识。

海　獭

海獭从观察周围环境开始一天的生活。海獭感到放心，知道自己在巨藻林中的作用。同时，她也完全清楚巨藻林在生态系统中扮演的角色。海獭并没有对为了获得皮毛而一直捕杀她的人类保持怨恨，似乎已经忘记和原谅了。海獭更愿意笑谈人类试图数她的毛发数量，而不是指责人类过去的错误。海獭被巨藻的友善和欣赏的语言所折服，于是也话语体贴。

艺术

The Arts

海藻广泛分布于全世界的海岸带。在大多数地区，它被视为杂草，应该扔进肥料堆。现在，你可以用它进行编织，将它转变为持久的材料。它在太阳下晒干后会变得坚硬、结实、耐磨，并适用于任何天气条件。

思维拓展

Systems: Making the Connections

生态系统是生命的网络。海岸带起到屏障的作用，保护陆地免受风暴潮、飓风和海啸损害。这片从海岸到海底的区域富含营养物质，因此是生物多样性的热点地区。巨藻林底层有海胆和海星栖居，海蟹和海龙也十分繁盛。巨藻的顶部是漂浮的，因为其具有气囊，使叶片能保持在水面，以便最大程度暴露在太阳下。一片陆地森林能支持3个门的动物生存，而巨藻林却至少能支持10个门的动物。依赖巨藻的物种如此多种多样，以至于查尔斯·达尔文断言："……我不认为其他哪种森林会像巨藻林一样，一旦毁灭就会引起那么多物种的消亡。"巨藻林还储存了数百万吨碳，要是这些碳被释放，会导致海洋酸化。因此，应对二氧化碳过度释放到大气中的一种方法同样是种植更多的巨藻。巨藻林有与热带雨林相近的生产力，但如果那些移动缓慢的鱼类和海洋无脊椎动物不受控制地生长，它们将影响巨藻林的生长和生态功能的发挥。

动手能力

Capacity to Implement

我们需要了解更多有关巨藻林的信息。首先，你可以与你的朋友、家人分享你学到的知识，即巨藻林和热带雨林同样重要。这种新的认识意味着你现在可以计划在南非和纳米比亚海岸的巨藻林中进行一次潜水之旅，即便不能真正成行，也要做这个计划。这个旅程将会让人惊叹，相当于参观亚马孙盆地。但为了享受这次旅程，你最好先学会如何潜水。

故事灵感来自

This Fable Is Inspired by

纳尔逊·曼德拉
Nelson Mandela

纳尔逊·曼德拉通过同他爱的人和最好的朋友一起手牵手，向全世界展示他的友爱和慈悲。当他离开监狱时，他牵起妻子的手，同时举起握紧的拳头。他永远不对逮捕他的人持怨恨态度，而是以尊重的方式对待他们，甚至偶尔与他们开玩笑，了解他们的家庭情况并记得所有细节。这传递了一个团结、呵护、亲密和信任的信息。即便西方文化嘲笑手牵手的行为，纳尔逊·曼德拉清楚地表明这种非洲传统并没有错，有错的是那些认为这种传统有错的人。同朋友和家人牵手的纳尔逊·曼德拉为自己是一名非洲人而感到骄傲，并执着于维护他从父母那里学到的传统。

图书在版编目(CIP)数据

冈特生态童书.第三辑修订版：全36册：汉英对照 / (比)冈特·鲍利著;(哥伦)凯瑟琳娜·巴赫绘；何家振等译.—上海：上海远东出版社，2022

书名原文：Gunter's Fables

ISBN 978-7-5476-1850-9

Ⅰ.①冈… Ⅱ.①冈… ②凯… ③何… Ⅲ.①生态环境-环境保护-儿童读物—汉、英 Ⅳ.①X171.1-49

中国版本图书馆CIP数据核字(2022)第163904号

著作权合同登记号图字09-2022-0637号

策　　划　张　蓉

责任编辑　祁东城

封面设计　魏　来李　廉

冈特生态童书

手牵手

[比]冈特·鲍利　著

[哥伦]凯瑟琳娜·巴赫　绘

唐继荣　译